Made in the USA
Columbia, SC
07 April 2021

35762776R00068